SCÉAL NA N-UIMHREACHA

THE NUMBER STORY

SMALL BOOK ONE

ENGLISH - IRISH GAELIC

Numbers Teach Children
Their Number Names

written and illustrated by

MISS ANNA

Early Reader Edition of *The Number Story 1*
Bronze Medal Winner, 2016 Wishing Shelf Book Award

Cover by | Lumpy Publishing
Layout by | Lumpy Publishing
Translated by Patchy O' Hatrick
Coloring by Jieeun Woo and Maria Mirabella

Library of Congress Control Number: 2018902040

Names: Miss Anna, author.
Title: Number story : numbers teach children their number names / Miss Anna.
Description: Portland, OR: Lumpy Publishing, 2018.
Identifiers: ISBN 978-1-945977-92-3| LCCN 2018902040
Summary: The pictures and rhymes present stories which introduce numbers 0-10.
Subjects: LCSH Numeration—English—Irish Gaelic--Pictorial works--Juvenile literature. | BISAC JUVENILE NONFICTION /
Languages: English—Irish Gaelic
Classification: LCC QA141.3 .M57 2018 | DDC 513—dc23

Publisher: Lumpy Publishing
Website: www.missannabooks.com
Email: missanna@missannabooks.com

Paperback: ISBN 978-1-945977-92-3
Printed in the U.S.A.   1 3 5 7 9 10 8 6 4 2

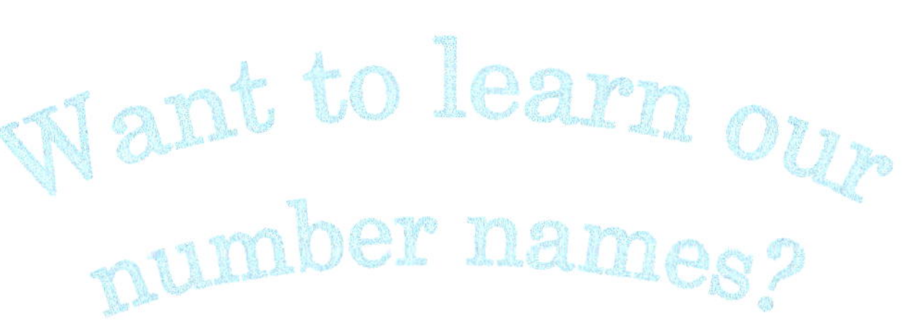

Ar mhaith leat ainmneacha
na n-uimhreacha a fhoghlaim?

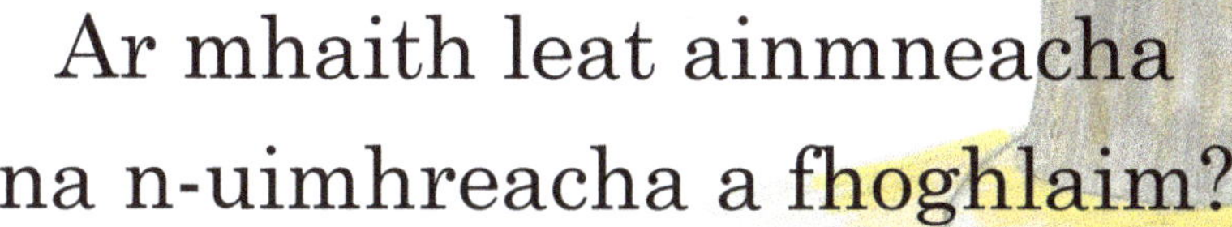

It is very easy and a lot of fun!

Is an-éasca agus an-chraic!

Say-along our little jingle

Abair in éindí linn ár scéilín!

starting from Number One!

Tpsóidh muid ón Uimhir a hAon!

1

ONE looks like my one finger.

A HAON

agus é cosúil le haon mhéar agam.

ONE!
A HAON!

2

TWO   trails a tail.

A DÓ

agus tarraingíonn sé eireaball.

A TAIL! EIREABALL!

THREE   has bumps.

A TRÍ

agus cnapáin air.

BUMPY! CNAPACH!

4

FOUR carries a sail.

A CEATHAIR agus seol air.

A SAIL!
SEOL!

5

FIVE   is a racing track.

A CÚIG

agus is ráschúrsa é.

VROOM
VRÚÚM!
1

SIX   curves like a snail.

A SÉ

agus é lúbach ar nós seilide.

A SNAIL! SEILIDE!

7

SEVEN has a sharp angle.

A SEACHT

agus uillinn ghéar air.

BE CAREFUL! IT'S SHARP!

Bí cúramach! Is Géar é!

8

EIGHT is rollercoaster rails.

A HOCHT

agus is ráillí ar roithleáin é.

HURÁ!
YIPPEE!

NINE is a bubble on a stick.

A NAOI

agus is boilgeog ar mhaide é.

A BUBBLE!

BOILGEOG!

TEN is an eye of a whale.

A DEICH

agus is leathshúil ar mhíol mór é.

WINK!
CAOCHADH!
HELLO! HÓIGH!

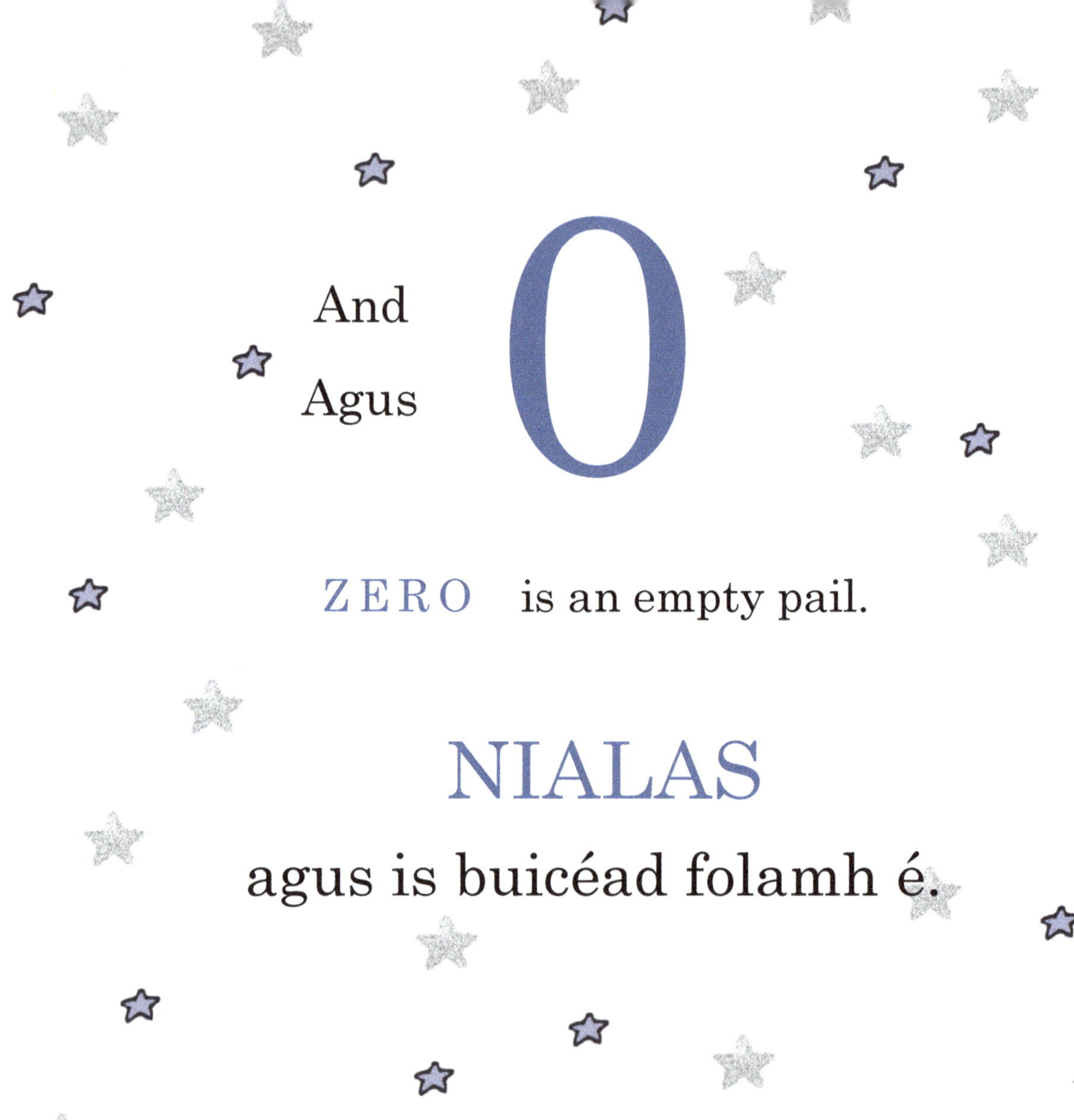

And
Agus

O

ZERO   is an empty pail.

NIALAS

agus is buicéad folamh é.

IT'S EMPTY!
Tá sé Folamh!

Thank you for playing with us today.

We had a lot of fun too!

Gabh raibh maith agat as imirt linn inniu.

Bhí an-chraic againn freisin!

We are your Number friends,
Zero to Ten,
Who will be here for you~
Is muidne do chairde Uimhreacha
Nialas go dtí an Deich.
Beidh muid i gcónaí anseo duit!

Bye-bye now!
See you again soon!
Slán leat anois!
Feicfidh muid a chéile gan mhoill!

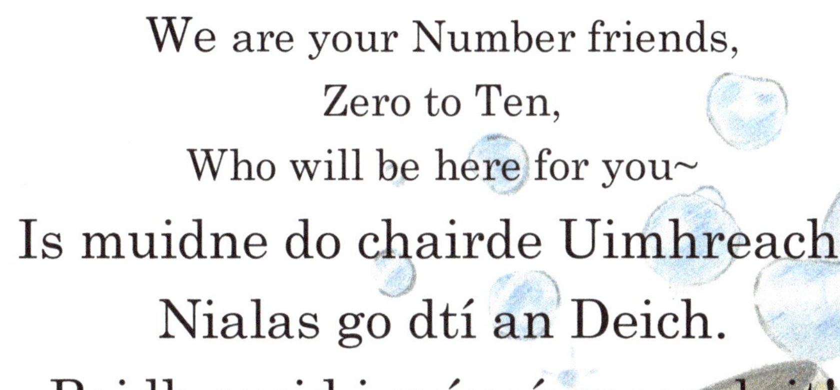

The Numbers are *SINGING* too!

To sing-a-long, look for Miss Anna Number Story
at your favorite music store like iTUNES.

MP3

Numbers 0-10
IDENTIFYING
& COUNTING

Numbers 11-20
& Ordinals
first, second, third...

Numbers 0-100
& Place Values
ones, tens, hundreds...

About Clock
& Telling Tim
hours, minutes, secor

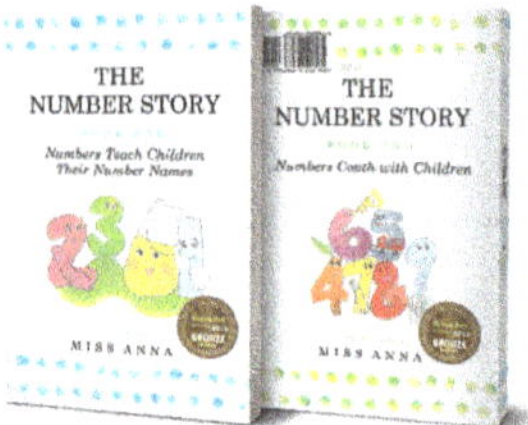

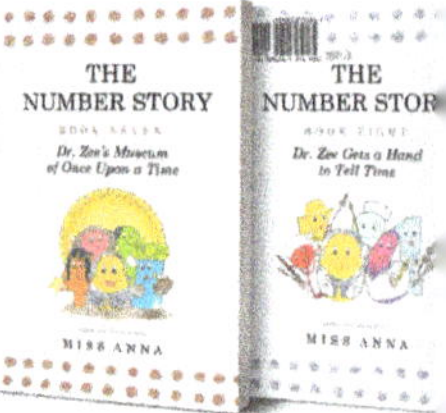

Number Story 1 & 2
isbn: 978-0-996216-48-7

Number Story 3 & 4
isbn: 978-1-945977-01-5

Number Story 5 & 6
isbn: 978-1-945977-06-0

Number Story 7 &
isbn: 978-1-949320-4

For more Miss Anna books to love,
visit us at

www.missannabooks.com

Numbers are working hard all over the world!
Come Travel the World with Us!

9 781945 977923